CATERPILLAR MILITARY TRACTORS
VOLUME 2 PHOTO ARCHIVE

CATERPILLAR MILITARY TRACTORS

VOLUME 2 PHOTO ARCHIVE

Photographs from the
Caterpillar Inc. Corporate Archives

Edited with introduction by
P. A. Letourneau

Iconografix
Photo Archive Series

Iconografix
P.O. Box 18433
Minneapolis, Minnesota 55418 USA

Text Copyright © 1994 by Iconografix

The photographs and illustrations are the property of Caterpillar Inc. and are reproduced with their permission.

All rights reserved. No part of this work may be reproduced or used in any form by any means…graphic, electronic, or mechanical, including photocopying, recording, taping, or any other information storage and retrieval system…without written permission of the publisher.

Library of Congress Card Number 94-76266

ISBN 1-882256-17-4

94 95 96 97 98 99 00 5 4 3 2 1

Cover and book design by Lou Gordon, Osceola, Wisconsin

Printed in the United States of America

Book trade distribution by Voyageur Press, Inc. (800) 888-9653

PREFACE

The histories of machines and mechanical gadgets are contained in the books, journals, correspondence and personal papers stored in libraries and archives throughout the world. Written in tens of languages, covering thousands of subjects, the stories are recorded in millions of words.

Words are powerful. Yet, the impact of a single image, a photograph or an illustration, often relates more than dozens of pages of text. Fortunately, many of the libraries and archives that house the words also preserve the images.

In the *Photo Archive Series*, Iconografix reproduces photographs and illustrations selected from public and private collections. The images are chosen to tell a story—to capture the character of their subject. Reproduced as found, they are accompanied by the captions made available by the archive.

The Iconografix *Photo Archive Series* is dedicated to young and old alike, the enthusiast, the collector and anyone who, like us, is fascinated by "things" mechanical.

ACKNOWLEDGMENTS

The photographs, captions, and related information included in this book were made available by the Caterpillar Inc. Corporate Archives. We are grateful to Caterpillar Inc. and to Joyce Luster, Corporate Archivist, for permission to reproduce the materials.

Caterpillar D6 with LaPlant-Choate bulldozer aids in landing equipment and supplies during the New Guinea Campaign. January 1944.

INTRODUCTION

Since the early years of World War I, Caterpillar tractors, engines, and power units have played critical roles in military campaigns around the globe. The first Caterpillar tractors to enter military service were sold by the Holt Manufacturing Company to the French War Department in 1914. Sales to the British and Russian governments followed in 1915. The Ordinance Division of the United States Army ordered its first Caterpillars in 1916. The tractors were used to move artillery and to transport ammunition and other supplies, previously the work of horses. Wartime demand for Caterpillars was such that the Holt factories were operated six days a week. By the end of World War I, the Stockton and Peoria plants had produced 5,082 tractors for the various military. Another 4,689 tractors were built by other manufacturers under Holt patent license and with the company's supervision.

Sales to the military diminished during the 1920s and 30s. The onset of World War II, however, brought a renewed demand. In 1940, 30 percent of Caterpillar Tractor Company's production was shipped overseas for use in military construction. The United States' entry into the war escalated the demand for bulldozers, road graders, engines, and electric sets to the point that 85 percent of the company's output went directly to the government. Military sales during the Korean War were significant as well, and Caterpillar equipment played a vital role in support of the United Nations' actions.

Caterpillar Military Tractors Photo Archive, Volume 1 and *Volume 2* feature photographs of the Caterpillar machines that saw action in the United States' invasion of Mexico in 1916, World War I, World War II, and the Korean War. While many of the tractors were photographed on the battlefield or behind Allied lines, most were either photographed at the factory or during field trials carried out by the military. *Volume 1* includes a special section covering World War I-era Caterpillar self-propelled gun mounts. *Volume 2* includes a special section covering the Caterpillar Diesel Military Engine or RD-1820, the air-cooled radial engine developed by the company in 1941 for use in M4 tanks. Both volumes include reproductions of paintings commissioned by Caterpillar and used in the company's advertisements during World War II.

UNITED STATES INVASION OF MEXICO
WORLD WAR I

B.G. Boultwood's painting of 1918, *A Caterpillar in Albert.*

Caterpillar Tractor Company No. 1, United States Quarter Master Corps (USQMC). 1917.

USQMC Holt 75 with trailers operating at the Mexican border. 1917.

Hauling with a Holt 75. 9th Infantry. Laredo, Texas.

USQMC Holt 75 with Caterpillar trailers operating at the Mexican border. 1917.

Holt 18 used by the Peoria army recruiting station in 1917.

Two views of a Military Type 45. 1916.

Holt 75 loaded on a train for shipment to Europe.

Holt 120 dynamometer tests carried out by Hyatt Roller Bearing Company. 1918.

Model 45 during tests for the U.S. Army.

Model 45 turning with a long train of Troy trailers. USQMC trials at Fort Sam Houston, Texas. 1917.

Two views of a camouflaged Armoured 5 Ton.

Armoured 10 Ton, the Model 55 Artillery Tractor, on parade in New York City.

Holt 75 somewhere on the Western Front.

Holt 18 undergoing tests for the U.S. Army Engineers.

Holt 10 Ton undergoing tests for the U.S. Army.

Caterpillar 45 Model E HVS with winch and armour.

A visit to the Peoria Holt plant by the Automobile Committee, Imperial Japanese Army . 1919.

A Holt 5 Ton at the top of Pikes Peak following a U.S. Army sponsored hill climb.

A 1915 demonstration of the Model 45 at Fort Sill, Oklahoma.

Holt 120 crossing rail tracks.

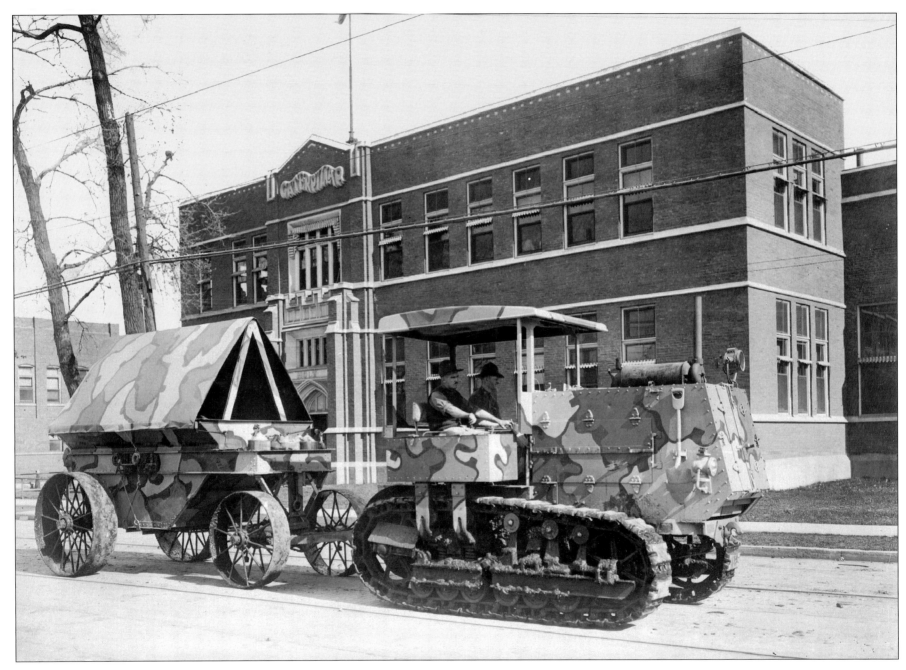

A camouflaged Armoured 10 Ton at the beginning of a non-stop drive from Peoria to Detroit.

Long Track 120 during 1918 U.S. Army tests.

Two views of a 1915 Model 18.

Armoured 10 Ton, the Model 55 Artillery Tractor.

Armoured 10 Ton, the Model 55 Artillery Tractor.

Front view of an Armoured 10 Ton.

Holt 10 Tons in France.

Model 75 with Troy trailers passing over very uneven ground. 1917.

Caterpillar 120 hauling artillery during New York Victory Parade. April 1919.

San Jose, California
March 1, 1955

The Caterpillar Tractor Co.

Dear Sirs-

Having read somewhere that you are celebrating your Golden Anniversary in business, I am enclosing a picture taken in France during World War I.

To give you some history of these old Holts, I drove one of these tractors as a member of the 62nd (heavy) Artillery Regiment. We had eight of these Holts in our regiment, all with friction drive. Four of them, I believe had six-cylinder motors, and the other four had four-cylinder motors. They had the tremendous speed of three miles per hour in high and about one mile per hour in low. They had what we called a "bull" wheel in front, and on a heavy pull the front would rear up like a rodeo bronco. It was quite tricky bringing her down again without jarring your teeth out.

Wishing you much continued success in your fine business, I am

Sincerely,

Herbert E. Bartholdi

Long Track 45 in creek test for U.S. Army. 1917.

Long Track 45 in creek test for U.S. Army. 1917.

Right side view of an Armoured 5 Ton at the factory.

The 81st Division of the Allied Expeditionary Forces in France.

Holt 18 hauling pontoon wagons across a bridge too weak for larger tractors. U.S. Engineering Tests. Leo Springs, Texas.

Holt 120 pulling a howitzer across battle terrain. 1918.

A right side view of a 5 Ton Artillery tractor.

Camouflaged 10 Ton under testing for the U.S. Army.

Armoured 10 Ton, the Model 55 Artillery Tractor.

Armoured 10 Ton, the Model 55 Artillery Tractor.

Camouflaged 10 Ton under testing for the U.S. Army.

A 1915 demonstration of the Model 45 at Fort Sill, Oklahoma.

Model 60 somewhere in England. 1915.

Holt 75 and Marine Corps personnel at Camp Le Jeune, Quantico, Virginia, at the time that the Marines were considering motorization of their heavy artillery. 1917.

Holt 75 with train of Caterpillar Wagons. Texas. 1917.

Three Holt 75s of the Tractor Section, Camp Joseph E. Johnston, Florida. November 8, 1918.

13th Field Army Artillery in Germany.

Long Track 120 during 1918 U.S. Army trails.

Holt 18 undergoing tests for the U.S. Army. Peoria 1918.

British War Department in France. 1915.

Holt 5 Ton.

Armoured 10 Ton, the Model 55 Artillery Tractor.

Model 45 E HVS. El Paso, Texas. 1917.

Caterpillar 75 pulling a big gun with the U.S. Army in France.

Holt 120 in Medina River during U.S. Army tests. The engine is protected from water thrown by the flywheel.

Gun Section, 18th Field Artillery, 3rd Division going into action.

Armoured 5 Ton with Holt trailer.

Testing a 5 Ton Artillery tractor and Caterpillar trailers.

Model 45 Artillery Tractor. May 18, 1917.

Testing the climbing characteristics of a 45 Long Track. 1917.

A Holt 120 with 4-truck trailer at Rhode Island Arsenal.

Canadian forces with a Holt 75 in Palestine.

Holt 18 of the 308th Field Signal Battery, Allied Expeditionary Force.

Somewhere in France.

WORLD WAR II
KOREAN WAR

Armed body guard for Caterpillar D4 of the Engineer Combat Regiment on Munda Airfield, New Georgia Island.

Before the Last Bomb Falls, the Lewis Daniel painting that illustrated Caterpillar advertisements of August and September 1942.

Coming at You, Schiklgruber! The painting that illustrated Caterpillar advertisements of June 1943. Artist unknown.

Wolf Hunt in the Sky, the Stahl painting that illustrated Caterpillar advertisements of September and October 1942.

Caterpillar D7 with LeTourneau angledozer working in an effort to get trucks through the Volturno Sector in Italy. November 1944.

Caterpillar D8s with LeTourneau Carryall scrapers building a base on Adak Island, Aleutian Islands. December 1943.

Caterpillar Diesel No. 12 Motor Graders leveling and grading an airfield somewhere in North Africa. January 1943.

Caterpillar D4 with LeTourneau bulldozer filling in tank trap dug by Italians on the road to Agrigento, Sicily. July 1943.

Four-Day Miracle on Adak, the Stevan Dohanos painting that illustrated Caterpillar advertisements of December 1942.

Caterpillar D4 at Atta Island, Aleutian Islands.

Caterpillar D7 drawing a 155mm gun at Borinquen Field, Fort Buchanan, San Juan, Puerto Rico. February 1941.

Caterpillar D8 with LeTourneau Carryall scraper re-leveling a landing field. March Field, Riverside, California. September 1941.

Untitled. Artist Stevan Dohanos.

Untitled. Artist Bingham.

Untitled. Artist unknown.

Jungle Fighter, the painting that illustrated Caterpillar advertisements of June and July 1942. Artist unknown.

Air Lightning Strikes from the Ground, the Noel Siklas painting that illustrated Caterpillar advertisements of November 1943.

Caterpillar D4 with LeTourneau sheep's foot roller, and No. 12 Motor Grader building a new air strip somewhere in New Guinea. February 1943.

Caterpillar D8 with LeTourneau bulldozer constructing an operational ramp on the island of Attu. September 1943.

Workpower is on Our Side, a painting that illustrated Caterpillar advertisements of 1944. Artist unknown.

Caterpillar D7 with blade grader at work on Rendova Island, Solomons. 1943.

Caterpillar D8 with bulldozer building a ramp for heavy equipment to come ashore on Dutch New Guinea. June 1944.

Caterpillar tractors unloading supplies from barges at Amchitka in the Aleutians. July 1943.

Title unknown. Artist Bingham.

Caterpillar D7 with LaPlant-Choate bulldozer clearing London streets of debris after an air raid. August 1942.

Road to Tokyo, the painting that illustrated Caterpillar advertisements of October and November 1942. Artist unknown.

U.S. engineers dismantling and demolishing a pontoon bridge across the Han River near Seoul, Korea. January 4, 1951.

Bulldozing over the sand bag foundation of an earth bridge for tank use, parallel to a lighter capacity pontoon bridge, on a tributary of the Naktang River, at the town of Shun-Shu, Korea.

Caterpillar Diesel D7 equipped with No. 7S bulldozer wading into the surf to extend line from its winch to anchor the causeway, at amphibious landing demonstrations at the Naval Amphibious Base, Little Creek, Virginia. September 1950.

A U.S. Army sergeant washes his socks in his helmet at the end of the day's maneuvers.

In July 1941, Caterpillar Tractor Co. was approached by the Ordinance Department to develop, for powering M4 tanks, a diesel radial engine of greater horsepower and wider adaptability than any then available.

By January 1942, Caterpillar engineers had produced a prototype based on a Curtiss-Wright aircraft engine. The 9-cylinder, air-cooled, radial design operated on a wide range of fuels, from crude oil to gasoline, without adjustment.

In February 1942, Caterpillar formed the Caterpillar Military Engine Company to build 1,000 engines per month. However, by late 1943, the Army's demand for Caterpillar bulldozers proved so critical that the government suspended production of the RD-1820 engine in favor of expanded production of the D7.

Only 120 RD-1820 engines were built. The following account of its development is reproduced from original documents preserved in the Caterpillar Inc. Corporate Archives.

The Challenge

The Ordinance Department has asked for a tank engine that will stand up under **battle conditions**, whether the battle is fought in the desert heat of North Africa or the sub-zero cold of Siberia. In describing battle conditions the Ordinance Department has depicted such adverse operating conditions that to be able to withstand them would seem almost beyond the endurance of man-made machinery. The job, therefore, becomes more than that of designing an engine to meet specifications; it becomes a challenge to the mechanical genius of American engineers.

Caterpillar engineers accepted the challenge because they believed they were particularly suited to design such an engine due to the resources available to them, and the experience they have had in designing engines to meet very similar operating conditions. For more than ten years, Caterpillar Diesel Engines, in track-type tractors, shovels and other types of heavy-duty equipment, have been operating successfully in dust, mud, sand, heat, cold, snow and ice; in foreign lands, on foreign fuels, with foreign operators and mechanics.

Development Background

To design a radial-type Diesel engine specifically for tank service is a major undertaking; and when time is a factor the job assumes huge proportions.

To eliminate as much preliminary work as possible and save time, a radial-type engine of proven qualities was needed for development work. No radial-type diesel engine could be found to meet Caterpillar requirements so a Wright Cyclone engine was chosen as the basis for this work. The wisdom of this choice has since been proved many times, not only from the standpoint of this modern engine's excellent design and construction, but also by the fine cooperation of the Wright Aeronautical Corporation.

Caterpillar development work proceeded. The following partial list of projects gives a general idea of the work involved:

Combustion chamber and pre-combustion chamber design and development.

Fuel injection system design and development.

Fuel injection system construction and testing.

Piston cooling design factors.

Exhaust and inlet cam timing and valve lift characteristics.

Cooling air flow and fan characteristics.

Surface finishes. Metallurgical and heat treatment problems.

Lubricants and break-in oil developments.

Lubricating oil temperature control.

Dust and dirt protection.

Starting system.

Accessories.

Instrumentation.

This work, performed by specialists in the respective fields, progressed along several different lines simultaneously. Tests were run continuously to prove design and check performance under laboratory duplicated battle conditions. As a result, the Caterpillar Diesel Military Engine is a highly perfected machine and is not expected to need "refining".

During the development stages it was found that many Wright parts, such as master rods, articulated rods, knuckle pins, supercharger and others, were entirely satisfactory for use in the Diesel engine. This would seem a definite advantage from a military standpoint, since many American planes are being powered with Wright engines. Consequently the transportation and supply of parts and maintenance requirements would be simplified for both plane and tank engines.

Forward view of the Diesel Military Engine on a test stand.

The Caterpillar Diesel Military Engine

The engine which Caterpillar offers the Ordinance Department is a supercharged, 4-stroke cycle, air-cooled, radial-type Diesel engine designed specifically for the job of powering tanks which must do battle in any part of the world. It provides the following general advantages:

Burns successfully a wide range of fuels—from residual type Diesel fuels to a mixture of 100 octane gasoline and lubricating oil. The widely distributed foreign Diesel fuels, such as Balik Papen from Borneo, give excellent performance.

Fuel system requires no operating adjustments and needs no timing, calibrating, setting or other adjustments when replacement parts are installed.

Gives tank a great cruising radius.

Uses the lubricating oils that meet U.S. Army specification No. 2-104A.

Idles smoothly for unlimited periods and has outstanding fuel economy at idling speed.

Operates dependably throughout a wide range of ambient air temperatures—minus 40 degrees Fahrenheit to hottest known desert temperatures.

Efficient dust and dirt protection.

Uses fuel which is free from vapor explosions.

Super lugging ability.

Upper and lower rpm limits give tank a speed range of 5 to 12-1/2 mph in third gear for best battle maneuverability.

Eliminates water supply problem for cooling.

Ample power and ruggedness with minimum weight.

No magnesium and very little aluminum (critical materials) required in its construction.

Simple in design to be easily understood from an operation and maintenance standpoint.

Engineered to operate at least 500 hours under severe battle conditions without any mechanical adjustments whatsoever.

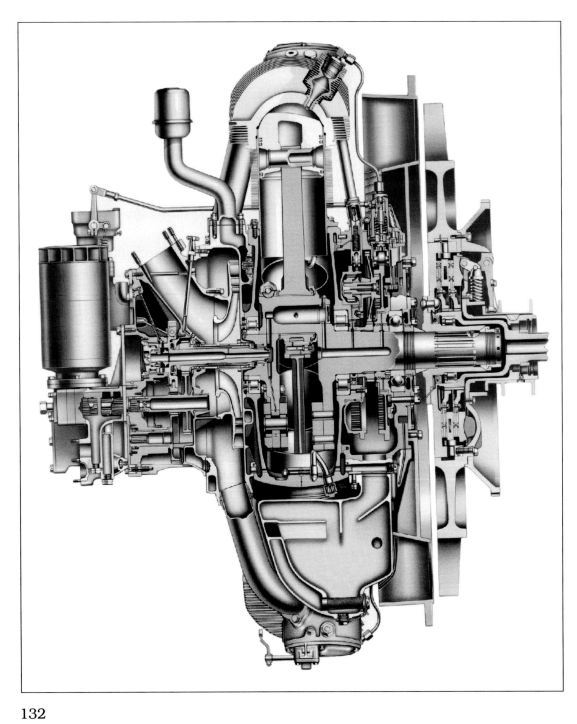

Cutaway view from the side.

Specifications of Caterpillar Diesel Military Engine (8-13-42)

Type	Radial Diesel
Horsepower at 2000 rpm under standard conditions with all accessories (including cooling system, but without transfer case)	450
Torque at engine crankshaft:	
At maximum horsepower	1,180 lb. ft.
Maximum at 1,200 rpm when engine is lugged down	1,470 lb. ft.
Torque at driveshaft:	
At maximum horsepower	746 lb. ft.
Maximum at 1,200 rpm when engine is lugged down	931 lb. ft.
Number of cylinders	Nine
Bore and stroke	6-1/8 x 6-7/8 in.
Piston displacement, total cu. in.	1,823
Stroke cycle	Four
Firing order	1-3-5-7-9-2-4-6-8
Piston speed; at 2000 rpm	2,292 fpm
Low idle speed, rpm	800
Compression ratio	15.5 to 1
Supercharger ratio	10 to 1
Overall dimensions, inches (approximate):	
Length, including generator and transfer case, but not exhaust pipes	56
Diameter	55
Weight (approximate), including clutch and transfer case, lb	3,500

Cutaway forward view.

Fuel—Burns all commercially available fuel oils, including fuels in accordance with U.S. Army Specification 2-102b, Class A; also all motor and aviation gasolines with sufficient lubricating oil added to assure engine ignition.

Fuel Injection System—Caterpillar designed and built. Gear-type transfer pump supplies fuel through two filters (cleanable metal edge type and replaceable absorbent type) to individual fuel injection pumps which deliver fuel to single orifice fuel injection valves. System is self-venting. Fuel injection pumps and fuel injection valves require no operating or installation adjustments.

Lubrication System—Full pressure, dry sump type. Two gear-type pumps. The scavenge pump delivers oil from sump through air-cooled oil coolers. Oil filters are full-flow metal edge types, plus replaceable bypass absorbent type. Lubrication oil—regular lubricating oils which meet U.S. Army specification 2-104-A.

Cooling System—Air-cooled by Caterpillar designed fan.

Starting System—Electric. Manifold air and engine compartment heaters for starting at sub-zero temperatures. Air Cleaners—Two 14-inch diameter Donaldson oil bath type.

Governor—Woodward hydraulic. Acts throughout entire engine speed range.

Pistons—Aluminum alloy. Cooled by constant jet of temperature controlled lubricating oil.

Cases and Housings—Main crankcase, crankcase front section, front and rear supercharger housings and cover of Caterpillar high strength cast iron.

Rendering from the forward side.

Rendering from the flywheel side.

Flywheel with fan.

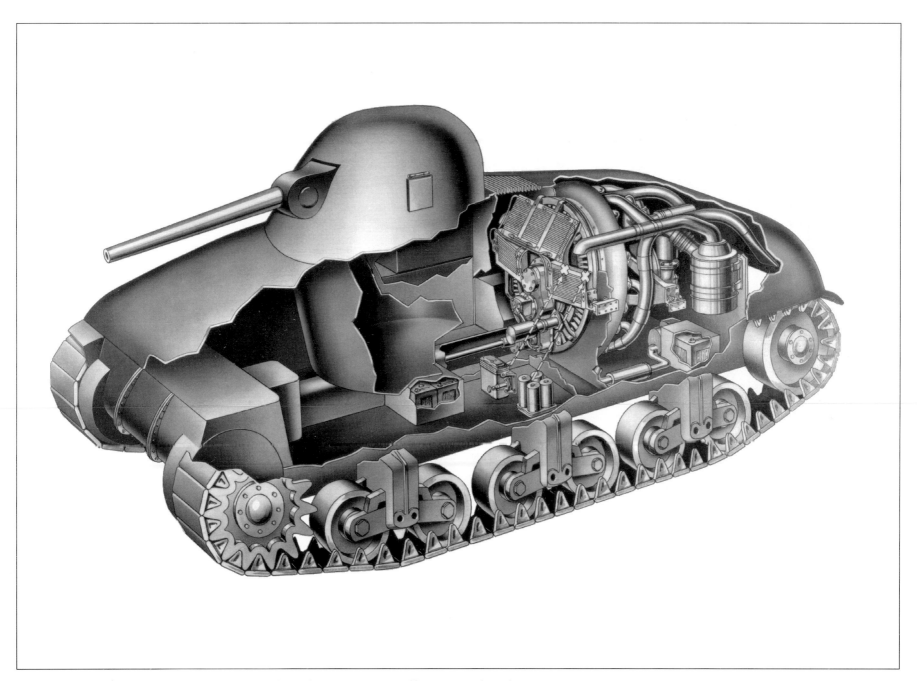

Rendering of an M4 tank equipped with the Caterpillar Diesel Military Engine.

CATERPILLAR TRACTORS IN WORLD WAR I
Built by Holt Manufacturing Company

	18 H.P.	45 H.P.	55 H.P.	60 H.P.	75 H.P.	120 H.P.	TOTAL
United States Government	2	74	2102	0	370	433	2081
British War Department	1	1	1	43	1362	243	1051
French War Department	0	352	0	0	18	0	370
Russian Government	0	0	0	20	60	0	80
Total	**3**	**407**	**2103**	**63**	**1810**	**676**	**5082**

Built for the U.S. Government by Other Manufacturers under Holt Patent License and Holt Supervision

Reo Motor Car Company	5 Ton Tractors	1477
Maxwell Motor Car Company	6 Ton Tractors (Renault Tanks)	225
	5 Ton Tractors	2193
Federal Motor Truck Company	2 1/2 Ton Tractors	87
Interstate Motor Company	2 1/2 Ton Tractors	7
Chandler Motor Car Company	10 Ton Tractors	700
Total		**4689**

The Iconografix Photo Archive Series includes:

JOHN DEERE MODEL D Photo Archive	ISBN 1-882256-00-X
JOHN DEERE MODEL A Photo Archive	ISBN 1-882256-12-3
JOHN DEERE MODEL B Photo Archive	ISBN 1-882256-01-8
JOHN DEERE 30 SERIES Photo Archive	ISBN 1-882256-13-1
FARMALL REGULAR Photo Archive	ISBN 1-882256-14-X
FARMALL F-SERIES Photo Archive	ISBN 1-882256-02-6
FARMALL MODEL H Photo Archive	ISBN 1-882256-03-4
FARMALL MODEL M Photo Archive	ISBN 1-882256-15-8
CATERPILLAR THIRTY Photo Archive	ISBN 1-882256-04-2
CATERPILLAR SIXTY Photo Archive	ISBN 1-882256-05-0
TWIN CITY TRACTOR Photo Archive	ISBN 1-882256-06-9
MINNEAPOLIS-MOLINE U-SERIES Photo Archive	ISBN 1-882256-07-7
HART-PARR Photo Archive	ISBN 1-882256-08-5
OLIVER TRACTOR Photo Archive	ISBN 1-882256-09-3
HOLT TRACTORS Photo Archive	ISBN 1-882256-10-7
RUSSELL GRADERS Photo Archive	ISBN 1-882256-11-5
MACK MODEL AB Photo Archive	ISBN 1-882256-18-2
MACK MODEL B, 1953-66 Photo Archive	ISBN 1-882256-19-0
CATERPILLAR MILITARY TRACTORS VOLUME 1 Photo Archive	ISBN 1-882256-16-6
CATERPILLAR MILITARY TRACTORS VOLUME 2 Photo Archive	ISBN 1-882256-17-4
LE MANS 1950: THE BRIGGS CUNNINGHAM CAMPAIGN Photo Archive	ISBN 1-882256-21-2
SEBRING 12-HOUR RACE 1970 Photo Archive	ISBN 1-882256-20-4
IMPERIAL 1955-1963 Photo Archive	ISBN 1-882256-22-0

The Iconografix Photo Archive Series is available from direct mail specialty book dealers and bookstores throughout the world, or can be ordered from the publisher.

For information write to:

Iconografix
P.O. Box 609
Osceola, Wisconsin 54020 USA

Telephone: (715) 294-2792
(800) 289-3504 (USA and Canada)
Fax: (715) 294-3414